ON THE ZOOLOGICAL

GEOGRAPHY

OF THE MALAY ARCHIPELAGO

(1859)

BY

ALFRED RUSSEL WALLACE

British Library Cataloguing-in-Publication Data
A catalogue record for this book is available from the
British Library

Alfred Russel Wallace

Alfred Russel Wallace was born on 8th January 1823 in the village of Llanbadoc, in Monmouthshire, Wales.

At the age of five, Wallace's family moved to Hertford where he later enrolled at Hertford Grammar School. He was educated there until financial difficulties forced his family to withdraw him in 1836. He then boarded with his older brother John before becoming an apprentice to his eldest brother, William, a surveyor. He worked for William for six years until the business declined due to difficult economic conditions.

After a brief period of unemployment, he was hired as a master at the Collegiate School in Leicester to teach drawing, map-making, and surveying. During this time he met the entomologist Henry Bates who inspired Wallace to begin collecting insects. He and bates continued exchanging letters after Wallace left teaching to pursue his surveying career. They corresponded on prominent works of the time such as Charles Darwin's *The Voyage of the Beagle* (1839) and Robert Chamber's *Vestiges of the Natural History of Creation* (1844).

Wallace was inspired by the travelling naturalists of the day and decided to begin his exploration career collecting specimens in the Amazon rainforest. He explored the Rio Negra for four years, making notes on the peoples and languages he encountered as well as the geography, flora, and fauna. On his return voyage his ship, Helen, caught fire and he and the crew were stranded for ten days before being picked up by the Jordeson, a brig travelling from Cuba to London. All of his specimens aboard Helen had been lost.

After a brief stay in England he embarked on a journey to the Malay Archipelago (now Singapore, Malaysia, and Indonesia). During this eight year period he collected more than 126,000 specimens, several thousand of which represented new species to science. While travelling, Wallace refined his thoughts about evolution and in 1858 he outlined his theory of natural selection in an article he sent to Charles Darwin. This was published in the same year along with Darwin's own theory. Wallace eventually published an account of his travels *The Malay Archipelago* in 1869, and it became one of the most popular books of scientific exploration in the 19th century.

Upon his return to England, in 1862, Wallace became a staunch defender of Darwin's landmark work *On the Origin of Species* (1859). He wrote responses to those critical of the theory of natural selection, including 'Remarks on the Rev. S. Haughton's Paper on the Bee's Cell, And on the Origin of Species' (1863) and 'Creation by Law' (1867). The former of these was particularly pleasing to Darwin. Wallace also published important papers such as 'The Origin of Human Races and the Antiquity of Man Deduced from the Theory of 'Natural Selection" (1864) and books, including the much cited *Darwinism* (1889).

Wallace made a huge contribution to the natural sciences and he will continue to be remembered as one of the key figures in the development of evolutionary theory.

Wallace died on 7th November 1913 at the age of 90. He is buried in a small cemetery at Broadstone, Dorset, England.

ON THE ZOOLOGICAL GEOGRAPHY
OF THE MALAY ARCHIPELAGO
(1859)

In Mr. Sclater's paper[1] on the Geographical Distribution of Birds, read before the Linnean Society, and published in the 'Proceedings' for February 1858, he has pointed out that the western islands of the Archipelago belong to the Indian, and the eastern to the Australian region of Ornithology. My researches in these countries lead me to believe that the same division will hold good in every branch of Zoology; and the object of my present communication is to mark out the precise limits of each region, and to call attention to some inferences of great general importance as regards the study of the laws of organic distribution.

The Australian and Indian regions of Zoology are very strongly contrasted. In one the Marsupial order constitutes the great mass of the mammalia,--in the other not a solitary marsupial animal exists. Marsupials of at least two genera (*Cuscus* and *Belideus*) are found all over the Moluccas and in Celebes; but none have been detected in the adjacent islands of Java and Borneo. Of all the varied forms of *Quadrumana*, *Carnivora*, *Insectivora*, and *Ruminantia* which abound in the western half of the Archipelago, the only genera found in the Moluccas are *Paradoxurus* and *Cervus*. The *Sciuridæ*, so numerous in the western islands, are represented in Celebes by only two or three species, while not one is found further east. Birds furnish equally remarkable illustrations. The Australian region is the richest in the world in Parrots; the Asiatic is (of tropical regions) the poorest. Three entire families of the Psittacine order are peculiar to the former region, and two of them, the Cockatoos and the Lories, extend up to its extreme limits, without a solitary species passing into the Indian islands of the Archipelago. The genus *Palæornis* is, on the other hand, confined with equal strictness to the Indian region. In the Rasorial order, the *Phasianidæ* are Indian, the *Megapodiidæ* Australian; but in this case one species of each family just passes the limits into the adjacent region. The genus *Tropidorhynchus*, highly characteristic of the Australian region, and everywhere abundant as well in the Moluccas and New Guinea as in Australia, is quite unknown in Java and Borneo. On the other hand, the entire families of *Bucconidæ*, *Trogonidæ* and *Phyllornithidæ*, and the genera *Pericrocotus*, *Picnonotus*, *Trichophorus*, *Ixos*, in fact, almost all the vast family of Thrushes and a host of other genera, cease abruptly at the eastern side of Borneo, Java, and Bali. All these groups are *common birds* in the great Indian islands; they abound everywhere; they are the characteristic features of the ornithology; and it is most striking to a naturalist, on passing the narrow straits of Macassar and

Lombock, suddenly to miss them entirely, together with the *Quadrumana* and *Felidæ*, the *Insectivora* and *Rodentia*, whose varied species people the forests of Sumatra, Java, and Borneo.

To define exactly the limits of the two regions where they are (geographically) most intimately connected, I may mention that during a few days' stay in the island of Bali I found birds of the genera *Copsychus*, *Megalaima*, *Tiga*, *Ploceus*, and *Sturnopastor*, all characteristic of the Indian region and abundant in Malacca, Java, and Borneo; while on crossing over to Lombock, during three months collecting there, not one of them was ever seen; neither have they occurred in Celebes nor in any of the more eastern islands I have visited. Taking this in connexion with the fact of *Cacatua*, *Tropidorhynchus*, and *Megapodius* having their western limit in Lombock, we may consider it established that the Strait of Lombock (only 15 miles wide) marks the limits and abruptly separates two of the great Zoological regions of the globe. The Philippine Islands are in some respects of doubtful location, resembling and differing from both regions. They are deficient in the varied Mammals of Borneo, but they contain no Marsupials. The Psittaci are scarce, as in the Indian region; the Lories are altogether absent, but there is one representative of the Cockatoos. Woodpeckers, Trogons, and the genera *Ixos*, *Copsychus*, and *Ploceus* are highly characteristic of India. *Tanysiptera* and *Megapodius*, again, are Australian forms, but these seem represented by only solitary species. The islands possess also a few peculiar genera. We must on the whole place the Philippine Islands in the Indian region, but with the remark that they are deficient in some of its most striking features. They possess several isolated forms of the Australian region, but by no means sufficient to constitute a real transition thereto.

Leaving the Philippines out of the question for the present, the western and eastern islands of the Archipelago, as here divided, belong to regions more distinct and contrasted than any other of the great zoological divisions of the globe. South America and Africa, separated by the Atlantic, do not differ so widely as Asia and Australia: Asia with its abundance and variety of large Mammals and no Marsupials, and Australia with scarcely anything but Marsupials; Asia with its gorgeous *Phasianidæ*, Australia with its dull-coloured *Megapodiidæ*; Asia the poorest tropical region in Parrots, Australia the richest: and all these striking characteristics are almost unimpaired at the very limits of their respective districts; so that in a few hours we may experience an amount of zoological difference which only weeks or even months of travel will give us in any other part of the world!

Moreover there is nothing in the aspect or physical character of the islands to lead us to expect such a difference; their physical and geological differences do not coincide with the zoological differences. There is a striking homogeneity in the two *halves* of the Archipelago. The great volcanic chain runs through both parts; Borneo is the counterpart of New Guinea; the Philippines closely resemble the equally fertile and equally volcanic Moluccas; while in eastern Java begins to be felt the more arid climate of Timor and Australia. But these resemblances are accompanied by an extreme zoological diversity, the Asiatic and Australian regions finding in Borneo and New Guinea respectively their highest development.

But it may be said: "The separation between these two regions is not so absolute. There *is* some transition. There *are* species and genera common to the eastern and western islands." This is true, yet (in my opinion) proves no transition in the proper sense of the word; and the nature and amount of the resemblance only shows more strongly the absolute and original distinctness of the two divisions. The exception here clearly proves the rule.

Let us investigate these cases of supposed transition. In the western islands almost the only instance of a group peculiar to Australia and the eastern islands is the *Megapodius* in North-west Borneo. Not one of the Australian forms of Mammalia passes the limits of the region. On the other hand, Quadrumana occur in Celebes, Batchian, Lombock, and perhaps Timor; Deer have reached Celebes, Timor, Buru, Ceram, and Gilolo, but not New Guinea; Pigs have extended to New Guinea, probably the true eastern limit of the genus *Sus*; Squirrels are found in Celebes, Lombock,

11

and Sumbawa: among birds, *Gallus* occurs in Celebes and Sumbawa, Woodpeckers reach Celebes, and Hornbills extend to the North-west of New Guinea. These cases of identity or resemblance in the animals of the two regions we may group into three classes; 1st, identical species; 2nd, closely allied or representative species; and 3rd, species of peculiar and isolated genera. The common Grey Monkey (*Macacus cynomolgus*) has reached Lombock, and perhaps Timor, but not Celebes. The Deer of the Moluccas seems to be a variety of the *Cervus rufus* of Java and Borneo. The Jungle Cock of Celebes and Lombock is a Javanese species. *Hirundo javanica, Zosterops flavus, Halcyon collaris, Eurystomus gularis, Macropygia phasianella, Merops javanicus, Anthreptes lepida, Ptilonopus melanocephala*, and some other birds appear the same in the adjacent islands of the eastern and western divisions, and some of them range over the whole Archipelago. But after reading Lyell on the various modes of dispersion of animals, and looking at the proximity of the islands, we shall feel astonished, not at such an amount of interchange of species (most of which are birds of great powers of flight), but rather that in the course of ages a much greater and almost complete fusion has not taken place. Were the Atlantic gradually to narrow till only a strait of twenty miles separated Africa from South America, can we help believing that many birds and insects and some few mammals would soon be interchanged? But such interchange would be a fortuitous mixture of faunas essentially and absolutely dissimilar, not a natural and regular transition from one to the other. In like manner the cases of identical species in the eastern and western islands of the Archipelago are due to the gradual and accidental commingling of originally absolutely distinct faunas.

In our second class (representative species) we must place the Wild Pigs, which seem to be of distinct but closely allied species in each island; the Squirrels also of Celebes are of peculiar species, as are the Woodpeckers and Hornbills, and two Celebes birds of the Asiatic genera *Phænicophæus* and *Acridotheres*. Now these and a few more of like character are closely allied to other species inhabiting Java, Borneo, or the Philippines. We have only therefore to suppose that the species of the western passed over to the eastern islands at so remote a period as on one side or the other to have become extinct, and to have been replaced by an allied form, and we shall have produced exactly the state of things now existing. Such extinction and such replacement we know has been continually going on. Such has been the regular course of nature for countless ages in every part of the earth of which we have geological records; and unless we are prepared to show that the Indo-Australian Archipelago was an altogether exceptional region, such must have been the course of nature here also. If these islands have existed in their present form only during one of the later divisions of the Tertiary period, and if interchange of species at very rare and distant intervals has occurred, then the fact of some identical and other closely allied species is a necessary result, even if the two regions in question had been originally peopled by absolutely distinct creations of organic beings, and there had never been any closer connexion between them than now exists. The occurrence of a limited number of representative species in the two divisions of the Archipelago does not therefore prove any true transition from one to the other.

The examples of our third class--of peculiar genera having little or no affinity with those of the adjacent islands--are almost entirely confined to Celebes, and render that island a district *per se*, in the highest degree interesting. *Cynopithecus*, a genus of Baboons, the extraordinary Babirusa and the singular ruminant *Anoa depressicornis* have nothing in common with Asiatic mammals, but seem more allied to those of Africa. A quadrumanous animal of the same genus (perhaps identical) occurs in the little island of Batchian, which forms the extreme eastern limit of the highest order of mammalia. An allied species is also said to exist in the Philippines. Now this occurrence of quadrumana in the Australian region proves nothing whatever as regards a transition to the western islands, which, among their numerous monkeys and apes, have nothing at all resembling them. The species of Celebes and Batchian have the high superorbital ridge, the long nasal bone, the dog-like figure, the minute erect tail, the predaceous habits and the fearless disposition of the true Baboons, and find their allies nowhere nearer than in tropical Africa. The *Anoa* seems also to point towards the same region, so rich in varied forms of Antelopes.

In the class of birds, Celebes possesses a peculiar genus of Parrots (*Prioniturus*), said to occur also in the Philippines; *Meropogon*, intermediate between an Indian and an African form of Bee-eaters; and the anomalous *Scissirostrum*, which Prince Bonaparte places next to a Madagascar bird, and forms a distinct subfamily for the reception of the two. Celebes also contains a species of *Coracias*, which is here quite out of its normal area, the genus being otherwise confined to Africa and continental India, not occurring in any other part of the Archipelago. The Celebes bird is placed, in Bonaparte's 'Conspectus,' between two African species, to which therefore I presume it is more nearly allied than to those of India. Having just received Mr. Smith's Catalogue of the Hymenoptera collected during my first residence in Celebes, I find in it some facts of an equally singular nature. Of 103 species, only 16 are known to inhabit any of the western islands of the Archipelago, while 18 are identical with species of continental India, China, and the Philippine Islands, two are stated to be identical with insects hitherto known only from tropical Africa, and another is said to be most closely allied to one from the Cape.

These phenomena of distribution are, I believe, the most anomalous yet known, and in fact altogether unique. I am aware of no other spot upon the earth which contains a number of species, in several distinct classes of animals, the nearest allies to which do not exist in any of the countries which on every side surround it, but which are to be found only in another primary division of the globe, separated from them all by a vast expanse of ocean. In no other case are the species of a genus or the genera of a family distributed in *two* distinct areas separated by countries in which they do not exist; so that it has come to be considered a law in geographical distribution, "that both species and groups inhabit continuous areas."

Facts such as these can only be explained by a bold acceptance of vast changes in the surface of the earth. They teach us that this Island of Celebes is more ancient than most of the islands now surrounding it, and obtained some part of its fauna before they came into existence. They point to the time when a great continent occupied a portion at least of what is now the Indian Ocean, of which the islands of Mauritius, Bourbon[2], &c. may be fragments, while the Chagos Bank and the Keeling Atolls indicate its former extension eastward to the vicinity of what is now the Malayan Archipelago. The Celebes group remains the last eastern fragment of this now submerged land, or of some of its adjacent islands, indicating its peculiar origin by its zoological isolation, and by still retaining a marked affinity with the African fauna.

The great Pacific continent, of which Australia and New Guinea are no doubt fragments, probably existed at a much earlier period, and extended as far westward as the Moluccas. The extension of Asia as far to the south and east as the Straits of Macassar and Lombock must have occurred subsequent to the submergence of both these great southern continents; and the breaking up and separation of the islands of Sumatra, Java, and Borneo has been the last great geological change these regions have undergone. That this has really taken place as here indicated, we think is proved by the following considerations. Not more than twenty (probably a smaller number) out of about one hundred land birds of Celebes at present known are found in Java or Borneo, and only one or two of twelve or fifteen Mammalia. Of the Mammalia and birds of Borneo, however, at least three-fourths, probably five-sixths, inhabit also Java, Sumatra, or the peninsula of Malacca. Now, looking at the direction of the Macassar Straits running nearly north and south, and remembering we are in the district of the monsoons, a steady south-east and north-west wind blowing alternately for about six months each, we shall at once see that Celebes is more favourably situated than any other island to receive stray passengers from Borneo, whether drifted across the sea or wafted through the air. The distance too is less than between any of the other large islands; there are no violent currents to neutralize the action of the winds; and numerous islets in mid-channel offer stations which might rescue many of the wanderers, and admit, after repose, of fresh migrations. Between Java and Borneo the width of sea is much greater, the intermediate islands are fewer, and the direction of the monsoons *along* and not *across* the Java sea, accompanied by alternating currents in the same direction, must render accidental communication between the two islands exceedingly difficult; so that where the facilities for intercommunication are greatest, the number of species common to the two countries is least, and *vice versâ*. But again, the mass of the species of Borneo, Java, &c., even when not *identical* are *congeneric*, which, as before explained, indicates *identity* at an earlier epoch; whereas the great mass of the fauna of Celebes is widely different from that of the western islands, consisting mostly of genera, and even of entire families, altogether foreign to them. This clearly points to a former total diversity of forms and species,-- existing similarities being the result of intermixture, the extreme facilities for which we have pointed out. In the case of the great western islands a former more complete identity is indicated, the present differences having arisen from their isolation during a considerable period, allowing time for that partial extinction and introduction of species which is the regular course of nature. If the very small number of western species in Celebes is all that the most favourable conditions for transmission could bring about, the complete similarity of the faunas of the western islands could never (with far less favourable conditions) have been produced by the same means. And what other means can we conceive but the former connexion of those islands with each other and with the continent of Asia?

In striking confirmation of this view we have physical evidence of a very interesting nature. These countries are in fact *still connected*, and that so completely that an elevation of only 300 feet would nearly double the extent of tropical Asia. Over the whole of the Java Sea, the Straits of Malacca, the Gulf of Siam, and the southern part of the China Sea, ships can anchor in less than fifty fathoms. A vast submarine plain unites together the apparently disjointed parts of the Indian zoological region, and abruptly terminates, exactly at its limits, in an unfathomable ocean. The deep sea of the Moluccas comes up to the very coasts of Northern Borneo, to the Strait of Lombock in the south, and to near the middle of the Strait of Macassar. May we not therefore from these facts very fairly conclude that, according to the system of alternate bands of elevation and depression that seems very generally to prevail, the last great rising movement of the volcanic range of Java and Sumatra was accompanied by the depression that now separates them from Borneo and from the continent?

It is worthy of remark that the various islands of the Moluccas, though generally divided by a less extent of sea, have fewer species in common; but the separating seas are in almost every case of immense depth, indicating that the separation took place at a much earlier period. The same principle is well illustrated by the distribution of the genus *Paradisea*, two species of which (the common Birds of Paradise) are found only in New Guinea and the islands of Aru, Mysol, Waigiou, and Jobie, all of which are connected with New Guinea by banks of soundings, while they do not extend to Ceram or the Ké Islands, which are no further from New Guinea, but are separated from it by deep sea. Again, the chain of small volcanic islands to the west of Gilolo, though divided by channels of only ten or fifteen miles wide, possess many distinct representative species of insects, and even, in some cases, of birds also. The Baboons of Batchian have not passed to Gilolo, a much larger island, only separated from it by a channel ten miles wide, and in one part almost blocked up with small islands.

Now looking at these phenomena of distribution, and especially at those presented by the fauna of Celebes, it appears to me that a much exaggerated effect, in producing the present distribution of animals, has been imputed to the accidental transmission of individuals across intervening seas; for we have here as it were a test or standard by which we may measure the possible effect due to these causes, and we find that, under conditions perhaps the most favourable that exist on the globe, the percentage of species derived from this source is extremely small. When my researches in the Archipelago are completed, I hope to be able to determine with some accuracy this numerical proportion in several cases; but in the mean time we will consider 20 per cent. as the probable maximum for birds and mammals which in Celebes have been derived from Borneo or Java.

Let us now apply this standard to the case of Great Britain and the Continent, in which the width of dividing sea and the extent of opposing coasts are nearly the same, but in which the species are almost all identical,--or to Ireland, more than 90 per cent. of whose species are British,--and we shall at once see that no theory of transmission across the present Straits is admissible, and shall be compelled to resort to the idea of a very recent separation (long since admitted), to account for these zoological phenomena.

It is, however, to the oceanic islands that we consider the application of this test of the most importance. Let any one try to realize the comparative facilities for the transmission of organized beings across the Strait of Macassar from Borneo to Celebes, and from South Europe or North Africa to the island of Madeira, at least four times the distance, and a mere point in the ocean, and he would probably consider that in a given period a hundred cases of transmission would be more likely to occur in the former case than one

in the latter. Yet of the comparatively rich insect-fauna of
Madeira, 40 per cent. are continental species; and of the
flowering plants more than 60 per cent. The Canary Islands
offer nearly similar results. Nothing but a former connexion
with the Continent will explain such an amount of specific
identity (the weight of which will be very much increased
if we take into account the representative species); and the
direction of the Atlas range towards Teneriffe, and of the
Sierra Nevada towards Madeira, are material indications of
such a connexion.

The Galapagos are no further from South America than
Madeira is from Europe, and, being of greater extent, are
far more liable to receive chance immigrants; yet they have
hardly a species identical with any inhabiting the American
continent. These islands therefore may well have originated in
mid-ocean; or if they ever were connected with the mainland,
it was at so distant a period that the natural extinction and
renewal of species has left not one in common. The character
of their fauna, however, is more what we should expect to
arise from the chance introduction of a very few species at
distant intervals; it is very poor; it contains but few genera,
and those scattered among unconnected families; its genera
often contain several closely allied species, indicating a single
antitype.

The fauna and flora of Madeira and of the Canaries, on
the other hand, have none of this chance character. They

are comparatively rich in genera and species; most of the principal groups and families are more or less represented; and, in fact, these islands do not differ materially, as to the general character of their animal and vegetable productions, from any isolated mountain in Europe or North Africa of about equal extent.

On exactly the same principles, the very large number of species of plants, insects, and birds, in Europe and North America, either absolutely identical or represented by very closely allied species, most assuredly indicates that some means of land communication in temperate or sub-arctic latitudes existed at no very distant geological epoch; and though many naturalists are inclined to regard all such views as vague and unprofitable speculations, we are convinced they will soon take their place among the legitimate deductions of science.

Geology can detect but a portion of the changes the surface of the earth has undergone. It can reveal the past history and mutations of what is now dry land; but the ocean tells nothing of her bygone history. Zoology and Botany here come to the aid of their sister science, and by means of the humble weeds and despised insects inhabiting its now distant shores, can discover some of those past changes which the ocean itself refuses to reveal. They can indicate, approximately at least, where and at what period former continents must have existed, from what countries islands must have been separated, and at how distant an epoch the rupture took place. By the invaluable indications which Mr. Darwin has deduced from the structure of coral reefs, by the surveys of the ocean-bed now in progress, and by a more extensive and detailed knowledge of the geographical distribution of animals and plants, the naturalist may soon hope to obtain some idea of the continents which have now disappeared beneath the ocean, and of the general distribution of land and sea at former geological epochs.

Most writers on geographical distribution have completely overlooked its connexion with well-established geological facts, and have thereby created difficulties where none exist. The peculiar and apparently endemic faunæ and floræ of the oceanic islands (such as the Galapagos and St. Helena) have been dwelt upon as something anomalous and inexplicable. It has been imagined that the more simple condition of such islands would be to have their productions identical with those of the nearest land, and that their actual condition is an incomprehensible mystery. The very reverse of this is however the case. We really require no speculative hypothesis, no new theory, to explain these phenomena; they are the logical results of well-known laws of nature. The regular and unceasing extinction of species, and their replacement by allied forms, is now no hypothesis, but an established fact; and it necessarily produces such peculiar faunæ and floræ in all but recently formed or newly disrupted islands, subject of course to more or less modification according to the facilities for the transmission of fresh species from adjacent continents. Such phenomena therefore are far from uncommon. Madagascar, Mauritius, the Moluccas, New Zealand, New Caledonia, the Pacific Islands, Juan Fernandez, the West India Islands, and many others, all present such peculiarities in greater or less development. It is the instances of identity of species in distant countries that presents the real difficulty. What was supposed to be the more normal state of things is really exceptional, and requires some hypothesis for its explanation. The phenomena of distribution in the Malay Archipelago, to which I have here called attention, teach us that, however narrow may be the strait separating an island from its continent, it is still an impassable barrier against the passage of any considerable number and variety of land animals; and that in all cases in which such islands possess a tolerably rich and varied fauna of species mostly identical, or closely allied with those of the adjacent country, we are forced to the conclusion that a geologically recent disruption has taken place. Great Britain, Ireland, Sicily, Sumatra, Java and Borneo, the Aru Islands, the Canaries and Madeira, are cases to which the reasoning is fully applicable.

In his introductory Essay on the Flora of New Zealand, Dr. Hooker[3] has most convincingly applied this principle to show the former connexion of New Zealand and other southern islands with the southern extremity of America; and I will take this opportunity of calling the attention of zoologists to the very satisfactory manner in which this view clears away many difficulties in the distribution of animals. The most obvious of these is the occurrence of Marsupials in America only, beyond the Australian region. They evidently entered by the same route as the plants of New Zealand and Tasmania which occur in South temperate America, but having greater powers of dispersion, a greater plasticity of organization, have extended themselves over the whole continent though with so few modifications of form and structure as to point to a unity of origin at a comparatively recent period. It is among insects, however, that the resemblances approach in number and degree to those exhibited by plants. Among Butterflies the beautiful *Heliconidæ* are strictly confined to South America, with the exception of a single genus (*Hamadryas*) found in the Australian region from New Zealand to New Guinea. In Coleoptera many families and genera are characteristic of the two countries; such are *Pseudomorphidæ* among the Geodephaga, *Lamprimidæ* and *Syndesidæ* among the Lucani, *Anoplognathidæ* among the Lamellicornes, *Stigmoderidæ* among the Buprestes, *Natalis* among the Cleridæ, besides a great number of representative genera. This peculiar distribution has hitherto only excited astonishment, and has confounded all ideas of unity in the distribution of organic beings; but we now see that they are in exact accordance with the phenomena presented by the flora of the same regions, as developed in the greatest detail by the researches of Dr. Hooker.

It is somewhat singular, however, that not one *identical species* of insect should yet have been discovered, while no less than 89 species of flowering plants are found both in New Zealand and South America. The relations of the animals and of the plants of these countries must necessarily depend on the same physical changes which the Southern hemisphere has undergone; and we are therefore led to conclude that insects are much less persistent in their specific forms than flowering plants, while among Mammalia and land birds (in which no genus even is common to the countries in question) species must die and be replaced much more rapidly than in either. And this is exactly in accordance with the fact (well established by geology) that at a time when the shells of the European seas were almost all identical with species now living, the European Mammalia were almost all different. The duration of life of species would seem to be in an inverse proportion to their complexity of organization and vital activity.

In the brief sketch I have now given of this interesting subject, such obvious and striking facts alone have been adduced as a traveller's note-book can supply. The argument must therefore lose much of its weight from the absence of detail and accumulated examples. There is, however, such a very general accordance in the phenomena of distribution as separately deduced from the various classes or kingdoms of the organic world, that whenever one class of animals or plants exhibits in a clearly marked manner certain relations between two countries, the other classes will certainly show similar ones, though it may be in a greater or a less degree. Birds and insects will teach us the same truths; and even animals and plants, though existing under such different conditions, and multiplied and dispersed by such a generally distinct process, will never give conflicting testimony, however much they may differ as regards the amount of relationship between distant regions indicated by them, and consequently notwithstanding the greater or less weight either may have in the determining of questions of this nature.

This is my apology for offering to the Linnean Society the present imperfect outline in anticipation of the more detailed proofs and illustrations which I hope to bring forward on a future occasion.

www.ingramcontent.com/pod-product-compliance
Lightning Source LLC
Chambersburg PA
CBHW022058190326
41520CB00008B/805